JACKIE LEE HOFER

ONE HARMONIC WHOLE

AF698651

THE SONG OF THE UNIVERSE

SunShine Press Publications

SunShine Press Publications, Inc.
PO Box 333
Hygiene, CO 80533-0333
Email: *sunshinepress@sunshinepress.com*

Cover and book design by Bob Schram of Bookends

FIRST EDITION

Publisher's Cataloging-in-Publication Data
(Provided by Quality Books, Inc.)

Hofer, Jackie Lee: 1935-
One harmonic whole : the song of the universe / Jackie Lee Hofer. — 1st ed.
p. cm.
Preassigned LCCN: 98-96005
ISBN: 0-9615743-9-9
1. Whole and parts (Psychology)—Quotations, maxims, etc. 2. Whole and parts (Philosophy)—Quotations, maxims, etc. 3. Spiritual life. I. Title.
BF202.H64 1998 147
QBI98-200

Printed in the United States of America
5 4 3 2 1

Printed on recycled, acid-free paper using soy-based ink

Dedicated

to

Jacqueline Felicia Hofer

THE PATH OF *the written word*

is to make you see.

—JOSEPH CONRAD

Foreword

There is a discovery that we each must make in our lives, if we are to live up to our human potential. This discovery is simple and available to each of us, all the time. It is the discovery that we live in a world which is made up of our concepts, our thoughts, our conditioning.

Our world is a world of thought which divides what is whole into two. This deeply imbedded division reflects itself in every word and symbol we use. The result is that we are constantly moving through a series of choices in our lives—the either/or divided way of thinking.

This divided viewpoint is destroying us. The division, sorting and prediction of the world by thought is indeed effective. It builds bridges, it cures diseases, it wins wars. It is a dominating force. But, this is precisely why it does not work. Domination is blind. There are people sleeping under the bridges that thought builds. There are new diseases replacing the ones that thought cures. And, the very winning of war means that war must be waged. This divided viewpoint cannot see our connection to the whole.

Thought, as expressed in language, has become the basis of our individual identities and our social realities. However, the whole is not in thought, it is not in language and it does not reside in any viewpoint or philosophy.

The whole is what we are. When we observe the division of thought and the illusion of opposites, in that moment the world becomes whole.

What will express itself from the whole? How will we live and relate? Can the totality of life, the undivided energy of consciousness infuse all and everything that we are? Can the world that thought built—the world of our own minds—suddenly become free from its divisions?

One Harmonic Whole points us in the direction of wholeness. Let us discover this in our lives. Let life discover us.

—STEVEN HARRISON
Boulder, Colorado

Preface

Ancient thinkers and modern scientists say that all things—living and nonliving—are interconnected, even to the entire universe. To become aware of this unbroken wholeness, we have to shift from a (black or white) divisive way of thinking to a (black and white) related way of thinking. Two short, simple words, *or/and*, can make a huge difference in how we view, think and act—the one separates while the other connects.

Some people, including many Native Americans, view life and the cosmos from an interconnected perspective where earth, wind, sky, water, rocks, animals and people are all related. The universe is the birthing source of all that was, is and ever will be. And, to that source all will return.

The purpose of this book is to help dissolve the barriers of racism, prejudice, violence, exclusiveness, intolerance, separateness and other limited ways of thinking. A related and connected way of thinking can have profound healing effects on everything from ecology to psychology and how we deal with all things—living and nonliving—and even with ourselves.

This book, like the symbol on the cover and the symbols throughout, points to an enfolding and unfolding of life in the universe. The connecting thread in this writing and the accompanying quotations is to show that we are indeed all part of *One Harmonic Whole.*

IF THE UNIVERSE IS TRULY INFINITE

in time and space,

then the implications go well beyond cosmology

to the whole view of nature,

to religion, philosophy, and society as a whole.

—ERIC LERNER

Read this book so that we can
individually and collectively
change our relatedness and connectedness
to life, society and the universe
by how we think, act and use our words,
labels, symbols, pictures and concepts.

We can easily

see that

the labels:

The unrelated

human being

lacks wholeness.

—Carl Jung

man

or

woman

IT IS TIME *to put our minds*
and hearts together as One
and realize that
"we all put our feet
against the same earth."

—RICHARD KASTL

black

or

white

THE DIVERSITY *in the human family*
should be the cause of love and harmony,
as it is in music where many different notes
blend together in the making of a perfect chord.

—ABDU'L-BAHÁ

american

or

russian

WE FEAR AND CRAVE

becoming

truly ourselves.

—HASIDIC SCRIPTURE

jew

or

christian

To make *an apple pie*

from scratch

you must first

invent the universe.

—Carl Sagan

salt

or

pepper

More and more

it comes to the fore that the only theory worthy

of our modern times (is to) forget literature—

politics and sociology must combine

all the bulkpeople of all lands,

the women not forgetting. . . .

Goodbye to all.

—Walt Whitman's last words

democrat

or

republican

The primary awareness

in ourselves, in animals and in primitives,

invariably consists of patterns of wholeness.

—Max Scheler

cat

or

dog

WORLDS COME AND GO *eternally.*

Then they are withdrawn again into the unmanifest alogical whole, only to be projected again; appearing and disappearing with the outbreathing and inbreathing power of all Creation, the pulse of life itself.

—FRANK WATERS

land

or

sea

True quiet means

keeping still when the time has come to keep still,

and going forward when the time has come to go forward.

In this way rest and movement are in agreement with

the demands of time, and thus there is light in life.

—I Ching

leftist

or

rightist

The Infinity of God

is neither person or a phenomenon;

it is a whole oneness, embracing everything—

every being, every phenomenon—

it is the endless universe itself.

—Michio Kushi

plant

or

animal

ONE TOUCH *of nature*

makes the whole world kin.

—WILLIAM SHAKESPEARE

rain

or

snow

MORE BROADLY ONE CAN SAY

that through the human being,

the universe is making a mirror to observe itself.

Or vice versa the universe could be regarded

as continuous with the body of the human being.

After all, this latter, like the plant,

gets all its substance and energy from the universe

and eventually falls back into it.

—David Bohm and B. J. Hiley

car

or

bicycle

I'VE SEEN *these beautiful patterns*
that the universe makes and
I'd like to know how it does it.

—MARGARET GELLER

star

or

galaxy

If I chase it,

I separate into the chaser and chased.

If I am it,

I am it and nothing else.

—John Lilly

are limited

and do not

include all things.

THE IRREVERSIBILITY (OF TIME)

is the mechanism that brings

order out of chaos.

—ILYA PRIGOGINE

Also, we can easily

see that

the labels:

ENHANCED AWARENESS *promotes in all of us*

the traits that abound in the creative person;

Whole-seeing.

Fresh, childlike perceptions.

Playfulness, a sense of flow.

Risk taking.

—MARILYN FERGUSON

male

or

female

I have a dream *today*!

—Martin Luther King, Jr.

african-american

or

caucasian

OUR BONDAGE IS NOT *failure to win our liberation but to see it. In fact, it is our aim that stands in the way of its realization; it is our anxiety to arrive at perfection in the unreal and unrealizable future which hides the Perfection which is of this very moment staring us in the face.*

—D. E. HARDING

capitalist

or

communist

THERE IS NO FEAR IN LOVE;

but perfect love casteth out fear because fear hath torment.

He that feareth is not made perfect in love.

—I JOHN

judaism

or

christianity

ONE COULD NOT *pluck a flower*

without disturbing a star.

—LOREN EISELEY

sodium chloride

or

piperaceae

DESCRIPTIONS OF UNITY *undoubtedly sound strange. This is because our language is adapted to frame either-or statements. We can say that things are many or that they are one, but what the mystic—the person in Unity—perceives is that they are* both *many* and *are one. This cannot be expressed in words without contradiction.*

—DR. ANTHONY CAMPBELL

liberal

or

conservative

We are "seeing," *just "seeing,"*

and what is seen is HERE,

always Here and always Now.

—Wei Wu Wei

feline

or

canine

All is beautiful,

all is beautiful,

all is beautiful, indeed.

—Navajo saying

soil

or

ocean

We are all *part of a spirit, a force, a* will, *which is irrepressible. It is towards this spirit, this unity, that we all strive. All humans, regardless of our faith or our political colorations, strive towards the eventual unity of the human spirit in eternity. We strive towards this, consciously or not. Some of us fail, some of us lean on others. Those of us who can perceive the paradox of our* aloneness *and yet at the same time our* unity *with the Whole can defeat fear. We can triumph over the worst death of all, the death of the human spirit!*

—Tristan Jones

radical

or

fundamentalist

Extinct

is forever.

—Kurt Benirschke

vegetable

or

vertebrate

LET GO OF *what you're holding on to,*

and everything that is really yours

will appear in your life

as if by magic.

—PAUL WILLIAMS

liquid

or

crystal

THE WORLD ONLY EXISTS

in your eyes—your conception of it.

You can make it as big

or as small as you want to.

—F. SCOTT FITZGERALD

automobile

or

velocipede

'TIS BUT A PART *we see,*

and not a whole.

—ALEXANDER POPE

sun

or

andromeda

IT IS ONLY WHEN *you seek it that you lose it.*

You cannot take hold of it, nor can you get rid of it;

while you can do neither, it goes on its own way.

You remain silent and it speaks;

you speak and it remains silent.

—ALAN WATTS

while more technical sounding

still do not

include all things.

Ultimately, humanity is one

and this small planet is our only home.

If we are truly to help one another

and protect this home of ours,

each of us needs to experience a vivid sense

of compassion and responsibility.

—The Dalai Lama

We can even see things as

symbols, pictures

and concepts:

When I ran away *from slavery,*
it was for myself;
when I advocated emancipation,
it was for my people;
but when I stood up for the rights of women,
self was out of the question.

—Fredrick Douglass

WE HUMANS HOLD *at least 99.9% of*

our DNA sequences in common.

We are far more closely related to one another

than we are to any other animal. . . .

Of the immense number of possible beings, realized and unrealized, we are all cut from the same cloth, made on the same pattern, granted the same strengths and weaknesses, and will ultimately share the same fate.

—CARL SAGAN AND ANN DRUYAN

I could have *gone on flying through space forever.*

—Yuri A. Gargarin

WHAT WE ARE LOOKING FOR

is what is looking.

—ST. FRANCIS OF ASSISI

It means *you can try to answer questions*

you thought the universe

was going to have to do without.

—Bruce Knapp

The true value of a human being *can be found in the degree to which he has found liberation from the self.*

—Albert Einstein

AMERICA HAS *tossed its cap over the wall of space.*

—JOHN F. KENNEDY

TEACH YOUR CHILDREN *what we have taught our children, that the earth is our mother. Whatever befalls the earth befalls the children of the earth. If we spit upon the ground, we spit upon ourselves. This we know. The earth does not belong to us; we belong to the earth.*

One thing we know, which the white man may one day discover, our God is the same God. You may think now that you own Him as you wish to own the land; but you cannot. He is the God of all people and His compassion is equal for all. The earth is precious to God, and to harm the earth is to heap contempt on its' creator.

So love it as we have loved it. Care for it as we have cared for it. And with all your mind, with all your heart, preserve it for your children, and love as God loves all.

—CHIEF SEATTLE

YOU ARE BOTH QUESTION AND ANSWER.

You are the total consciousness for which you seek,

but bound by life's many experiences.

One thing must be clear to you: there is no interruption;

the experience and you are the same.

—EILEEN GARRET

AND AS WE LET OUR OWN LIGHT SHINE,

we unconsciously give other people

permission to do the same.

As we are liberated from our own fear,

our presence automatically liberates others.

—NELSON MANDELLA

Art in everlasting change

. . . so is the whole universe.

—Marcus Aurelius Antoninus

MAN MAKES A GREAT FUSS

about this planet which is only

a ball-bearing in the hub of the universe . . .

—CHRISTOPHER MORLEY

IT SUGGESTS THAT THE UNIVERSE *is not a dead thing, composed of separate parts in mechanical interaction; rather, it is organic—an almost living thing in which each aspect subsists in some underlying activity. The more we give room to the subtle, the more we come to touch the heart of the universe and drink of its creative waters.*

—F. DAVID PEAT

Without the courage *to confront one's inner as well as one's outer worlds, such wholeness is impossible to achieve. Instrumental reason alone cannot lead to it.*

—Joseph Weisenbaum

while more graphic and visual,

still do not

include all things.

PERHAPS THE HEALING OF THE WORLD

rests on just this sort of shift in our way of seeing,

a coming to know that in our suffering and in our joy

we are connected to one another

with unbreakable and compelling human bonds.

—RACHEL NAOMI REMEM, M.D.

In time, some things change and
are thought of in new ways including
our previously assigned words,
labels, symbols, pictures and concepts:

WHAT IS IN ONE is in the whole,

and therefore, ultimately,

each soul is responsible for the whole world.

—GARY ZUKAV

some men become women

and

some women become men

Everything is inextricably interrelated:

We come to realize that we are responsible for everything we do, say, or think, responsible in fact for ourselves, everyone and everything else, and the entire universe.

—S. Rinpoche

some blacks lighten

and

some whites darken

For one priceless moment

in the whole history of man,

all of the people on this earth are truly one.

—Richard M. Nixon

some americans become russians

and

some russians become americans

THE UNITIES, SIR,

are a completeness—a kind of

universal dovetailness with regard to

place and time—a sort of general oneness . . .

—CHARLES DICKENS

some christians become jews

and

some jews become christians

HOW WE TAKE

a fragment for the whole,

and call the whole a part.

—OLIVER WENDELL HOLMES

peppers accent

and

salts spice

THE VILLAGE CAN NO LONGER
be defined as a place on the map,
or a list of people or organizations,
but its essence remains the same:
it is the network of values and relationships
that support and affect our lives.

—HILLARY RODHAM CLINTON

some democrats become republicans

and

some republicans become democrats

The universe *is popping all over the place.*

—Riccardo Giacconi

some cats eat dogfood

and

some dogs eat catfood

SOONER OR LATER *everyone of us*

breathes an atom that has been breathed before

by anyone you can think of that has lived before us—

Michelangelo or George Washington or Moses.

ALL SCIENCE *is the search*

for unity in hidden likenesses.

—JACOB BRONOWSKI

lands become seas

and

seas become lands

NO AMOUNT *of food and shelter*

will solve the human condition.

Our condition is a result of our view, our mind, our illusion.

Our responsibility to help is to relate.

In relationship, we discover that the divisions

that thought creates dissolve into wholeness.

—STEVEN HARRISON

some fundamentalists reform

and

some radicals moderate

THE REAL WORLD

is enough.

—GEORGE SANTAYANA

rains become snow

and

snows become rain

When you have seen
one ant, one bird, one tree,
you have not seen them all.

—Edward Wilson

some animals digest plants

and

some plants digest animals

TRUTH IS

a pathless land.

—J. KRISHNAMURTI

people sometimes trade bicycles for cars

and

trade cars for bicycles

The created work

is but a parenthesis in eternity.

—Sir Thomas Browne

stars are born in galaxies

and

galaxies create stars

I DON'T THINK THAT PHYSICS *will ever have an end. I think that the novelty of nature is such that its variety will be infinite—not just in changing forms but in the profundity of insight and the newness of ideas.*

—ISIDOR RABI

The main feature
of the universe and life
is change.

UNDERLYING THE ISOLATED WORLD *of ordinary objects*
and human experience is another reality,
an interconnected world of intermingling
relationships and possibilities.

—DEAN I. RADIN

All the millions of words, labels, symbols, pictures
and concepts we use are merely human ways
of identifying features of life in the universe.
They merely point to and describe things,
but are not the true things themselves.

In the end the aggressors

always destroy themselves, making way for others

who know how to cooperate and get along.

Life is much less a competitive struggle for survival

than a triumph of cooperation and creativity.

—Fritjof Capra

Instead of thinking and using

either/or

words, labels, symbols, pictures

and concepts we can think and use

and/and

words, labels, symbols, pictures

and concepts within any given category:

Social science affirms
that a woman's place in society
marks the level of civilization.

—Elizabeth Cady Stanton

Besides men and women,

there are boys and girls,

guys and gals,

masculine and feminine,

and many other words,

labels, symbols, pictures and concepts

to represent the sexes.

WE ARE A NATION OF

many nationalities, many races, many religions—

bound together by a single unity,

the unity of freedom and equality.

Whoever seeks to set one nationality against another,

seeks to degrade all nationalities.

Whoever seeks to set one race against another,

seeks to enslave all races.

Whoever seeks to set one religion against another,

seeks to destroy all religions.

—FRANKLIN D. ROOSEVELT

Besides the black and white races,

there are

brown, red and yellow races.

(Skin color is merely the human way of labeling races. Go to an art store, drugstore or school supply store and buy some colored construction paper. Take out one sheet each of black, brown, red, white, and yellow paper. Now, try to match these to any race and to anyone you know. They never match! This skin pigmentation is a result of the amount of melanin in the skin. These human designated colors are false and dangerous as they allow discrimination and control based on the amount of melanin in a person's skin. In truth, they are all part of the human race.)

THE EARTH *is but one country,*

and mankind its citizens.

—Bahá'u'lláh

Besides Americans and Russians,
there are Canadians and Australians and British and
Irish and Chinese and Japanese and Italians and
Egyptians and Eskimos and Finnish and Israelis and
Palestinians and Mexicans and Brazilians and Native
Americans and Africans and Swedish and Germans and
French and Spaniards and hundreds of other
nationalities and cultures.

We are all citizens

of one world, we are all of one blood.

To hate a man because he was born in another country,

because he speaks a different language,

or because he takes a different view on

this subject or that, is a great folly.

Desist, I implore you, for we are all equally human.

—Johann Amos Comenius

Besides Christians and Jews,

there are Baha'is and Buddhists and Hindus and

Mormons and Muslims and Seventh Day Adventists

and Zorasterians and hundreds of other

religious organizations.

By thoughts

I embrace the universal.

—Blaise Pascal

Besides salts and peppers,

there are herbs and spices

and hundreds of other seasonings.

A *PESSIMISTIC* OPTIMIST

is one who sadly shakes his head and says:

"I'm very much afraid that everything is for the best!"

—RAYMOND SMULLYAN

Besides democrats and republicans,
there are reform parties and independents and greens
and labour and libertarians and progressives
and moderates and hundreds of other
political parties throughout the world.

Without women,

we stood in space on one leg only.

—Vladimir Dzanibekovl

Besides cats and dogs,
there are horses and elephants and hippopotamuses
and lions and bears and gerbils and giraffes
and many other animals.

THE UNIVERSE

is but one

vast symbol of God.

—THOMAS CARLYLE

Besides lands and seas,

there are mountains and plains and rivers and volcanoes

and streams and hills and creeks and steppes and

lakes and deserts and ponds and many other

geographic features throughout the earth.

CONSERVATIVES AND LIBERALS *may be defined as those who think that on the whole things are all right and you better not tinker much with them, and those who think that there is enormous room for improvement.*

—JUSTICE FELIX FRANKFURTER

Besides leftists and rightists,

there are liberals and conservatives and radicals

and socialists and many other ideologies

throughout the world.

AT THE DEEPEST LEVEL,

our world can be regarded as a pattern

in infinite-dimensional space,

a space in which we and our minds

move like fish in water.

—RUDY RUCKER

Besides plants and animals,

there are insects and flowers and trees and birds

and fish and fungi and bacteria and

countless other organisms on earth.

THE WORLD

is its own magic.

—SHUNRYU SUZUKI

Besides rain and snow,
there are hail and fog and sleet and ice
and glaciers and wind and many other
climatic features on earth.

Change is necessary for life.

Everyone has a path.

No one's path is exactly the same.

I respect that truth.

—Peny Goodson-Kjome

Besides cars and bicycles,

there are trucks and motorcycles and buggies and planes

and busses and trains, and many other mechanical

means to move people on earth.

IF IT WERE NOT FOR SPACE

all matter would be jammed together in one lump

and that lump wouldn't take up any room.

—IRENE PETER

Besides stars and galaxies,
there are nebulas and meteors and comets
and moons and planets, and many other
astronomical features of the universe.

Human inquiry

can continue indefinitely

to yield important new truths.

—Henry Pierce Stapp

and on and on and on and on
and on and on and on and on
and on and on and on and on
and on and on.

WE MAY ALSO SEE

in the sky an aurora, which is cosmic plasma,

reminding us of the time when our world

was born out of plasma.

—HANNES ALFVÉN

The universe is made up of
the movement of living creatures
and nonliving things and things we haven't even
discovered and given words, labels, symbols,
pictures, concepts and names to.

ALL THIS UNIVERSE

is in this eye of mine.

—SEPPO

The universe is all these named
and unnamed things, and all
these named and
unnamed things are
the universe.

WE NOW HAVE THE WONDER,

not merely that we are all related to

and intimate with everything about us,

but that we have a cousin relationship

with every being in the universe,

especially with the living beings of planet earth.

—BRYAN SWIMM AND THOMAS BERRY

By looking at our world and our universe
we can see that in time everything changes.
We realize that we are all in this together
regardless of whatever words, labels, symbols,
pictures, concepts and names we think and use
in our everyday lives.

WE ARE ONE WORLD;

we share one earth, one water,

one fire, one sun.

—TRUNGPA RINPOCHE

The universe is a living presence
through which all things unfold and pass.
Everything is constantly changing and forever fresh.
We must honor, protect and celebrate
this diversity and unity of life
instead of fearing, conquering and destroying it.

WE MAY BE PART OF

an infinite and eternal multiverse within which new domains "sprout" into universes whose horizons never overlap . . .

—MARTIN REES

There are some theories, evidences and prophecies
that the universe is infinite—without beginning
and with no end in sight. Our universe,
the one we can see, may only be
one of many in a huge and infinite universe

Catch hold of the music

that lasts through eternity.

Search for the sun

that never sets.

—Mevlana Rumi

This unity and diversity
is the underlying reality
that permeates all things
from the smallest particle
to the largest supercluster.

STAND UNDER THE STARS

and say what you like to them.

Praise or blame them, question them,

pray to them, wish upon them.

The universe will not answer.

But it will have spoken.

—TIMOTHY FERRIS

We and all things living and nonliving,
named and unnamed, are the musical star dust
of a universal song.
Like the notes of a great symphony,
we are all part of One Harmonic Whole.
We join together with all things in the
celebration of the universal dance of life!

Reread this book so that we can
individually and collectively change our relatedness
and connectedness to life, society and the universe
by how we think, act and use our words, labels,
symbols, pictures and concepts.